THE GREAT
BARRIER
BEACH
FIELD GUIDE

THE GREAT BARRIER BEACH FIELD GUIDE

ANTHONY S MINARDI

ILLUSTRATIONS:
FRANCESCO BOLOGNA
MICHELE S. MOTT

EDITORS:
MICHAEL BUQUICCHIO
DEBRA ROSE

Copyright © 2017 by Anthony S Minardi.

Library of Congress Control Number:		2017913137
ISBN:	Hardcover	978-1-5434-4682-1
	Softcover	978-1-5434-4684-5
	eBook	978-1-5434-4683-8

All rights reserved. No part of this book may be reproduced or transmitted in any form or by any means, electronic or mechanical, including photocopying, recording, or by any information storage and retrieval system, without permission in writing from the copyright owner.

Any people depicted in stock imagery provided by Thinkstock are models, and such images are being used for illustrative purposes only.
Certain stock imagery © Thinkstock.

Print information available on the last page.

Rev. date: 11/10/2017

To order additional copies of this book, contact:
Xlibris
1-888-795-4274
www.Xlibris.com
Orders@Xlibris.com
758799

CONTENTS

Preface xiii

Surf Zone—Seaweeds 1

Surf Zone—Invertebrates 15

Introduction to Outer Face of Primary Dune 33

Swale Zone—Lee of Primary Dune 47

Bog Zone 67

Roadside—Borderline 77

References 91

Index 93

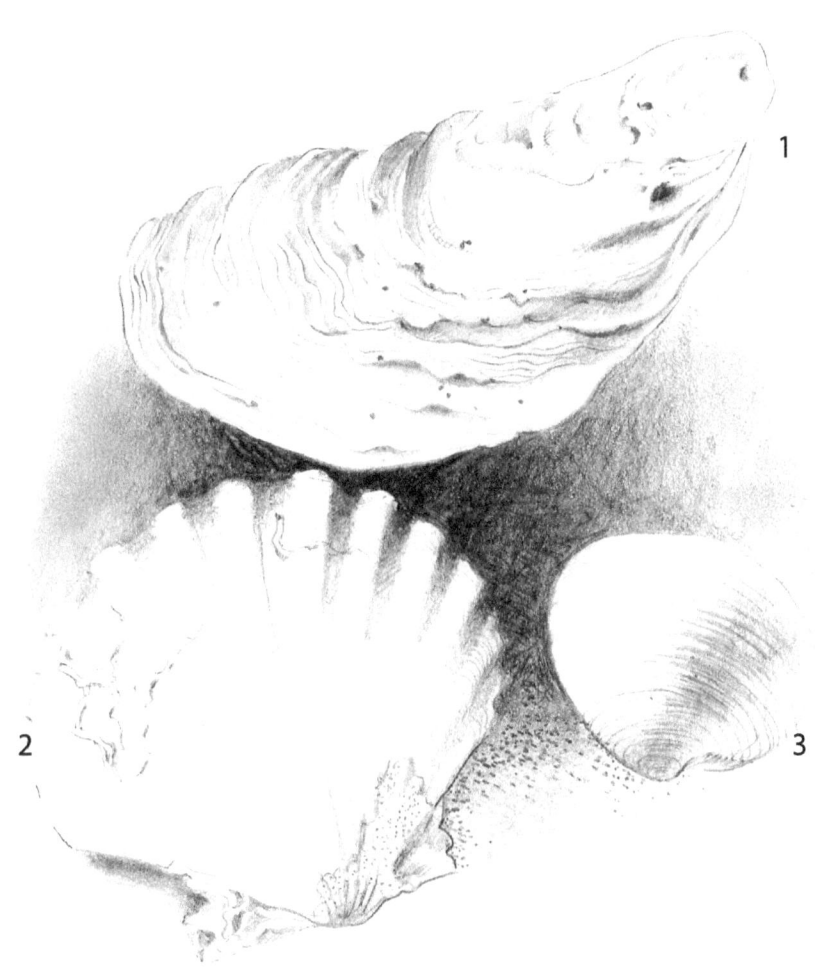

Hard Clam
Scallop
Eastern Oyster

A Biological Transect of The Barrier Beach

Procedure employed to flora and fauna distributions of the zones

Optional Procedure

Beach Transect

Materials

- Two five-foot wood stakes two-by-two-inch benchmarks
- Compass, camera, collecting container, clipboard, and pencil
- Fifty feet of clothesline rope, a roll of red ribbon
- One-square-meter wood frame

Procedure 1

- Calibrate the fifty-inch rope into ten-foot intervals; place a red ribbon at each interval.
- Place a benchmark at the upper limits of the primary dune.
- Place the second stake in line with the first stake at the low-tide mark. Each stake should be in line with each other. A compass reading is recommended to maintain position.
- Starting at stake 1 (primary dune), place the wood frame at the first ten square meters. Photograph the flora within the square. Clip a leaf and a floret from the plants, which will assist with identification.
- Repeat the same procedure every ten feet to the low-tide mark.
- You may repeat the process ten or twenty feet, east or west of your first transect, for an average of flora and fauna distribution

Procedure 2
(Lee Side of Primary Dune to High Ground)
(Zones: Swale, Bog, High Ground)

Procedure 2

- Starting from the lee side of the primary dune, place stake 1.
- Maintain a compass reading to high ground; place stake 2 (two benchmarks in line with each other).
- Starting at station 1 (lee side of primary dune), continue the process every ten feet to high ground.

You now have completed a biological transec tough ground. Using your clippings and photographs, proceed to organize your data and identify the samples observed. Reference materials, books, field guides are listed for genetic identification and general information as to their genetic adaptations for survival and perpetuation. You now have complete transec from Low tide to high ground. Computerise the total transec for future reference.

1 Introduction
2 References

Incorporate both transecs.

Preface

The publication *The Barrier Beach* is the result of several years of beach surveys of the Northeast Coast, Long Island, New York, New Jersey, New England, and the coastlines of North and South Carolina. "In its natural state, the barrier beach is a fascinating mixture of opposites: high and low elevations, wet and dry, hot and cold, sterile and fertile, windblown and sheltered." It is the combination of these factors that, with a strip of land varying in width from a few thousand feet to a few hundred or less, have produced several distinct zones of flora distribution. Considering the mixture of abiotic elements of the various zones, the floras that occupy a zone must be genetically equipped to adapt or be replaced by genetically equipped species. The environment is the selecting agent. It determines and selects the most genetically equipped to survive and perpetuate. I recommend that the reader take a few moments and review Darwin's theory of natural selection. The environment is in a state of changing abiotic factors and is selecting the genetic mutations that are required to survive and reproduce. The intent of this publication was to classify genetic classification. Most important are the genetic characteristics, mutations required to adapt to the abiotic conditions of the various zones of the barrier beach. The most important concept of the field guide is the zone the floras first inhabit, not to the zones they may extend to.

Surf Zone

Seaweeds

Identifications

02/27/17
Zone 1: Surf-Zone Seaweeds (Substrate *Mesoderma deauratum*)

Seaweeds:

Laminaria agardhii—kelp + substrate
Sargassum filipendula—gulfweed
Codium fragile—sponge seaweed
Fucus vesiculosus—rockweed
Chondrus crispus—Irish moss
Ascophyllum nodosum—knotted wrack
Rhodymenia palmata—dulse
Laminaria digitata—horsetails
Corallina officinalis—coral seaweed
Ulva lactuca—sea lettuce

Kelp

Kelp
Laminaria agardhii

Laminaria agardhii—kelp
Zone 1

Substrate
Sea clam
Spisula solidissima

Gulfweed
Sargassum fibipendula

Substrate
Sea clam

Substrate
Sea clam
Spisula solidissima

Sargassum filipendula—Gulfweed Zone 1
Sargassum, Cedar Point Gardiners Bay, September '80

Intertidal zone—surf zone
Zone 1

Codium fragile
Sponge seaweed

Substrate: whelk
Busycon canaliculatum

Codium fragile—sponge seaweed, Zone 1

Rockweed
Fucus vesiculosus

Fucus vesiculosus—rockweed
Zone 1

Chondrus crispus—Irish moss

Irish moss
Chondrus crispus

Ascophyllum nodosum—knotted wrack

Zone 1, berm of the beach
Zone 1

Dulse
Rhodymenia palmata + substrate *Spisula solidissma*

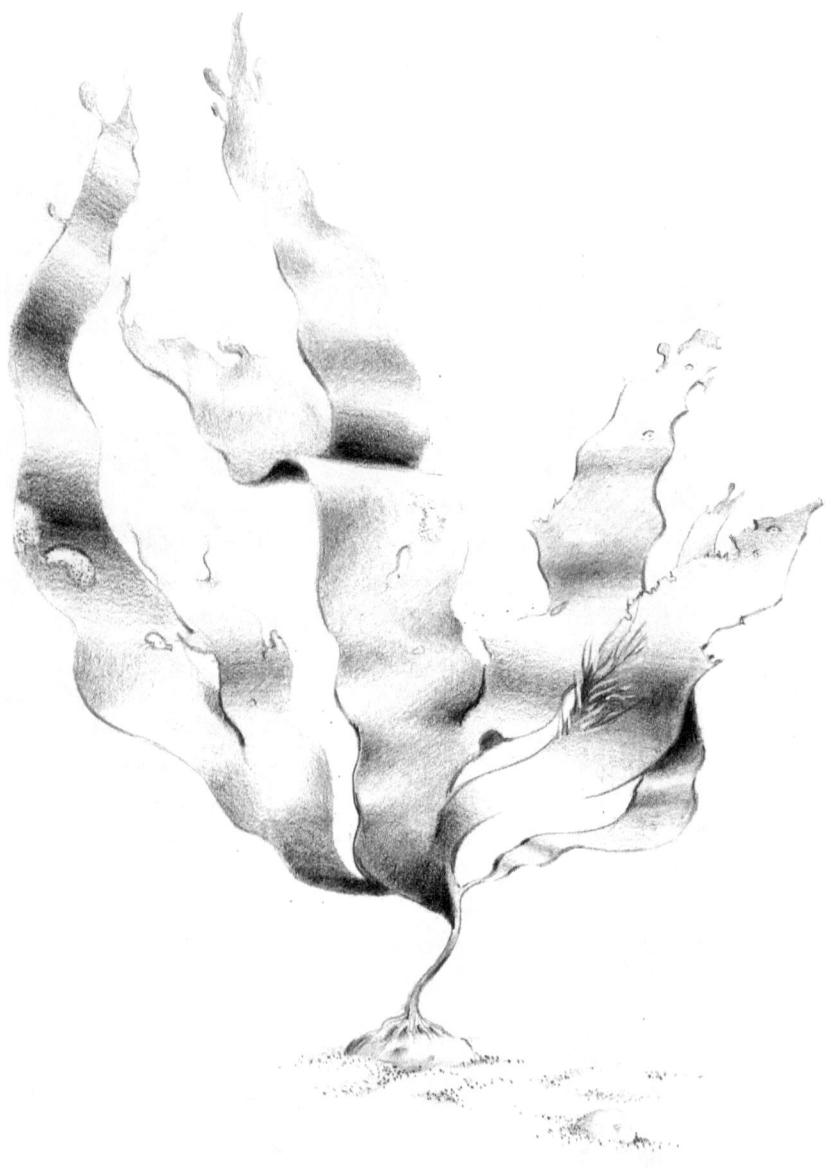

Rhodymenia palmata—dulse
Zone 1

Laminaria digitata—horsetails
Zone 1

Coral seaweed
Corallina officinalis

Corallina officinalis—coral Seaweed
Zone 1

Sea clam
Substrate
Spisula solidissima

Sea lettuce
Ulva lactuca

Ulva lactuca—sea lettuce
Spisula solidissima—sea clam
Zone 1

Surf Zone

Invertebrates

Twenty-three invertebrates identified drawings

Twenty-three plates attached

Tidal Zone

Intertidal to the Surf Zone

Final Index
For Surf Zone
Invertebrates
Surf Zone (Invertebrates)

Illustrations (Twenty-Three)

1. Surf clam—*Spisula solidissima*
2. Eastern oyster—*Ostrea virginca*
3. Common scallop—*Pecten irradians*
4. Sea clam (black clam)—*Arctica islandica*
5. Sand dollar—*Echinarachnius parma*
6. Sea scallop—*Pecten islandicus*
7. Jonah crab—*Cancer borealis*
8. Common periwinkle—*Littorina littorea* (hydroid attached)
9. Edible mussel—*Mytilus edulis* (barnacles attached)
10. Moon snail—*Polinices heros*
11. Sand collar—moon snail egg casing
12. Clamshell (acts as a substrate)—*Eupleura caudata*
13. Northern starfish—*Asterias forbesi*
14. Common ark shell—*Arcampechensis pexata*
15. Razor clam—*Ensis directus*
16. Jingle shells—*Anomia simplex*
17. Hard clam—*Mercenaria mercenaria*
18. Channeled whelk—*Busycon canaliculatum*
19. Skate egg casing—*Chondrichthyes*
20. Boring sponge—*Cliona celata*
21. Star coral—*Astrangia danae*
22. Horseshoe crab—*Limulus polyphemus*
23. Knobbed whelk—*Busycon carica*

Surf clam
Spisula solidissima

Plate 1

Ostrea virginca
Eastern oyster

Black clam
Pecten irradians

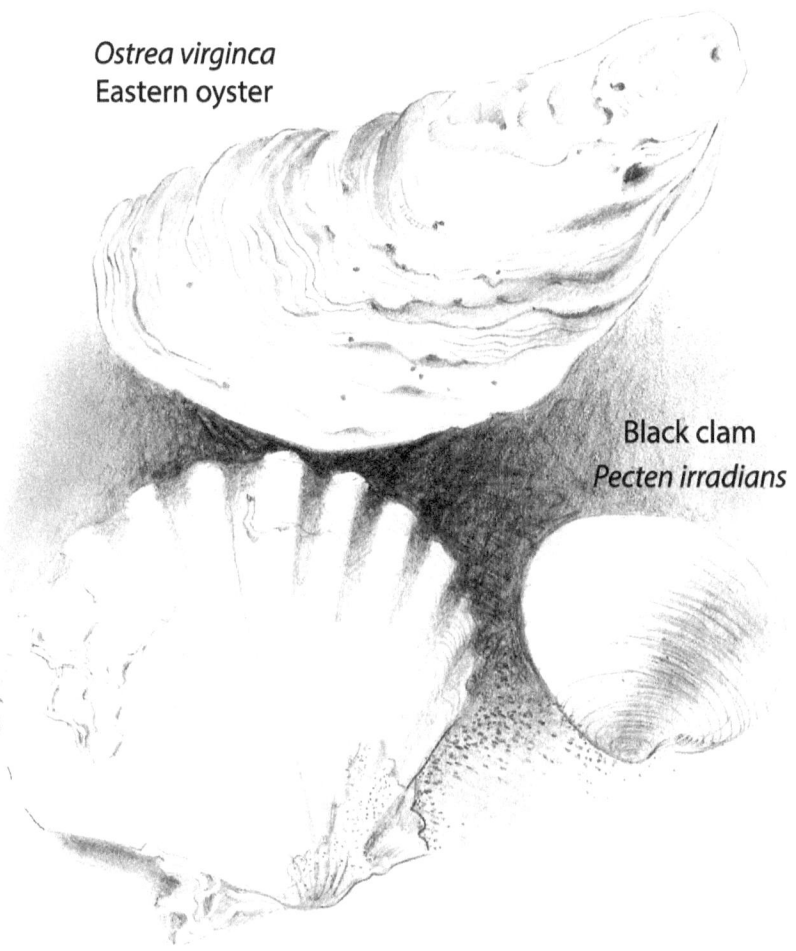

Common bay scallop
Arctica islandica

Plate 2

Sand dollar
Echinarachnius parma

Sea scallop
Pecten islandicus

Plate 3

Sand dollar, North Bar Montauk Point, September '79

Common snail

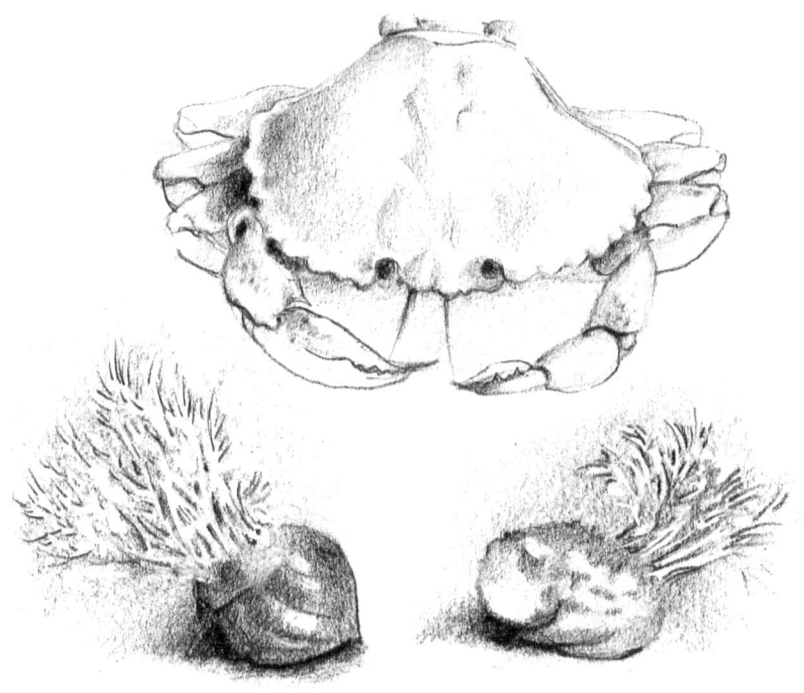

Jonah crab
Cancer borealis

Littorina littorea + Hydroid Attached
Hydroides species

Plate 4

Edible mussel
Mytilus edulis

Barnacles attached
Balanus balanoides

Plate 5

Moon snail egg casing

Moon snail
Polinices heros

Clamshell substrate
Eupleura caudata—oyster drill snail

Clamshell serves as a substrate for a snail—*Eupleura caudate* (snail) An oyster drill snail

Plate 6

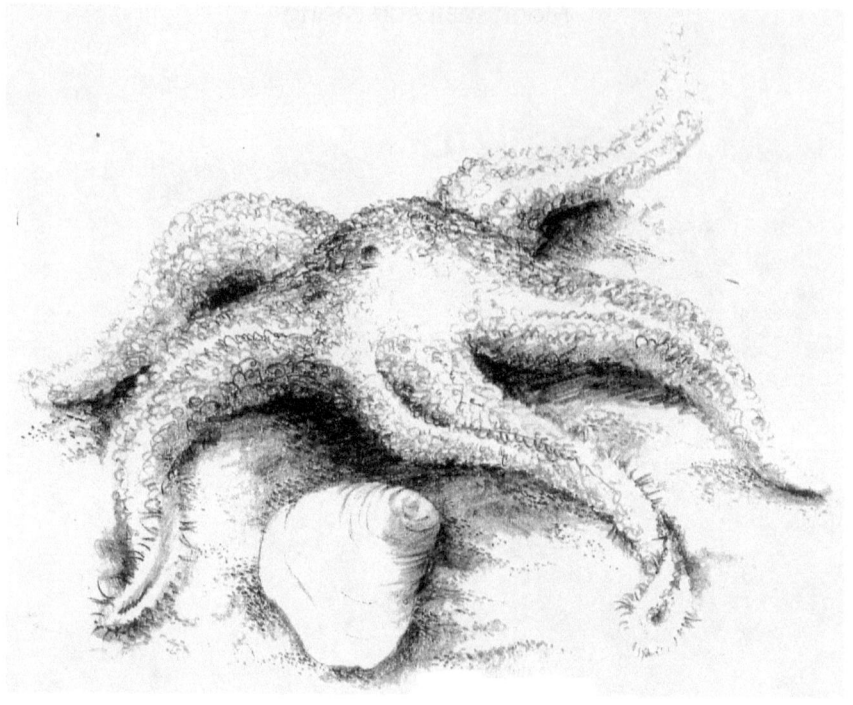

Northern starfish
Asterias forbesi

Plate 7

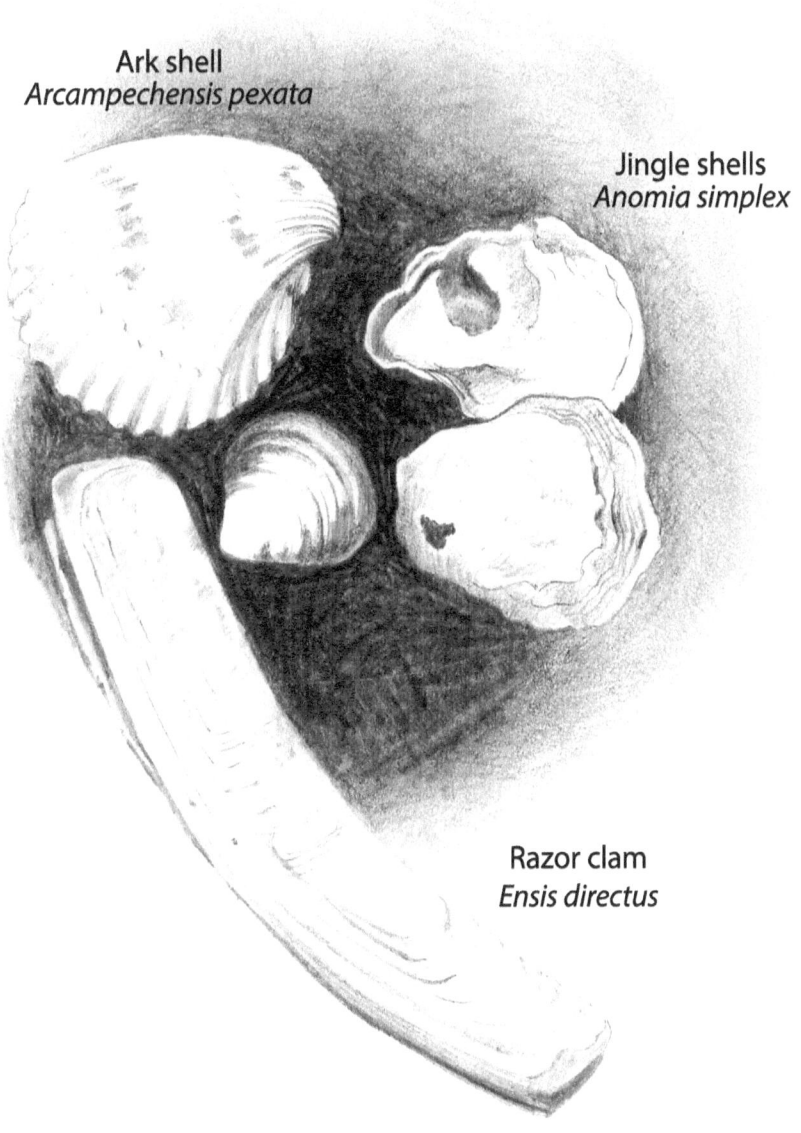

Plate 8

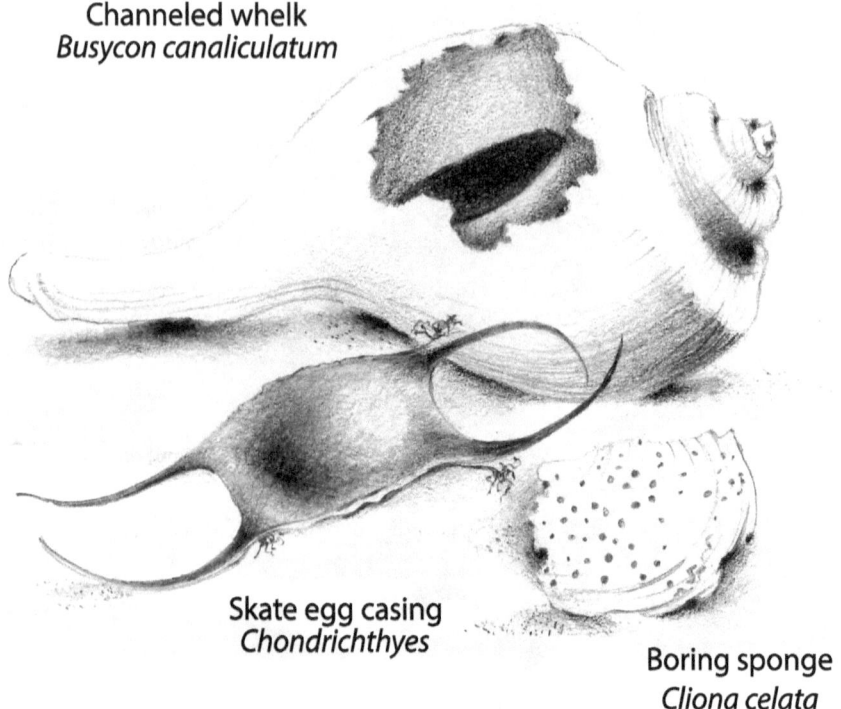

Star coral
Astrangia danae

Plate 10

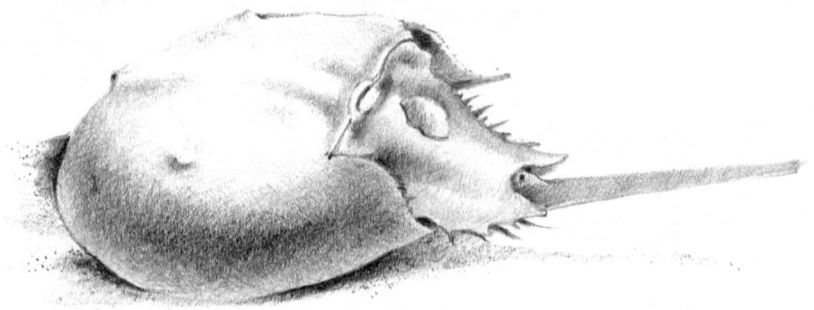

Horseshoe crab
Limulus polyphemus

Plate 11

Knobbed whelk
Busycon carica

Introduction to
Outer Face of Primary Dune

Windward Zone

Index—Final Sequence
04-11-17
Outer Face of the Primary Dune

1   Sea rocket—*Cakile edentula*
2   Common saltwort—*Salsola kali*
3   Seabeach sandwort—*Arenaria peploides*
4   Coast blite—*Chenopodium rubrum*
5   Coast blite—*Chenopodium icanium*
6   Beach grass—*Ammophila breviligulata*
7   Halberd-leaved orache—*Atriplex patula*
8   Dusty miller—*Artemisia stelleriana*
9   Cocklebur—*Xanthium echinatum*
10 Seaside goldenrod—*Solidago sempervirens*
11 Seaside spurge—*Euphorbia polygonifolia*

Sea rocket
Cakile edentula

Common saltwort
Salsola kali

Sandwort
Arenaria peploides

Coast blite
Chenopodium rubrum

Coast blite
Chenopodium icanium

Beach grass
Ammophila breviligulata

Halberd-leaved orache
Atriplex patula

Dusty miller
Artemisia stelleriana

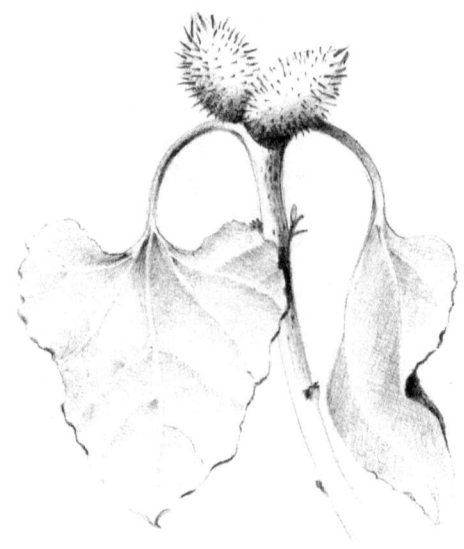

Cocklebur
Xanthium echinatum

Seaside Goldenrod
Solidago sempervirens

Seaside spurge
Euphorbia polygonifolia

Swale Zone

Lee of Primary Dune

Fifteen Illustrations

7

Swale
Lee of Primary Dune

Introduction to Swale Zone

1 Beach heath—*Hudsonia tomentosa*
2 Beach pea—*Lathyrus japonicus*
3 Common evening primrose—*Oenothera biennis*
4 Beach plum—*Prunus maritima*
5 Pinweed—*Lechea maritima*
6 Jointweed—*Polygonella articulata*
7 Sickle-leaved golden aster—*Chrysopsis falcata*
8 Prickly pear—*Opuntia humifusa*
9 Wormwood—*Artemisia caudata*
10 Bayberry—*Myrica pensylvanica*
11 Salt spray rose—*Rosa rugosa*
12 Lichen reindeer moss—*Cladonia tenuis*
13 Iceland moss—*Cetraria islandica*
14 British soldier lichen—*Cladonia cristatella*
15 Earthstar—*Geaster hygrometricus*

Intro to Swale Zone

Dusty Miller · Poison Ivy · Goldenrod

Beach heath
Hudsonia tomentosa

Beach pea
Lathyrus japonicus

Prim rose
Oenothera biennis

Beach plum
Prunus maritima

Pinweed
Lechea maritima

Jointweed
Polygonella articulata

Sickle-leaved golden aster
Chrysopsis falcata

Prickly pear
Opuntia humifusa

Wormwood
Artemisia caudata

Bayberry
Myrica pensylvanica

Salt spray rose
Rosa rugosa

Reindeer moss
Cladonia tenuis

Iceland moss (a lichen)
Cetraria islandica

British soldier lichen
Cladonia cristatella

Earthstar
Geaster hygrometricus

Bog Zone

8

Introduction to Bog Zone

Cranberry
Vaccinium macrocarpon

Bog Zone

Illustrations

Introduction Page to Bog Zone

1 Bearberry—*Arctostaphylos uva-ursi*
2 Bulrush—*Scirpus atrovirens*
3 Black grass—*Juncus gerardii*
4 Cranberry bog—*Vaccinium macarpon*
5 Seaside gerardia—*gerardia maritima*
6 Thread-leaved sundew—*Drosera filiformis*

Bearberry
Arctostaphylos uva-ursi

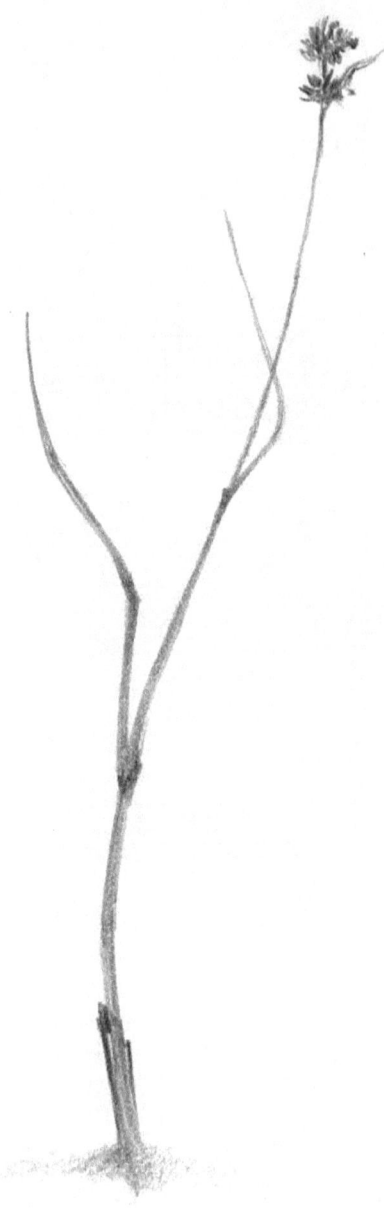

Bulrush
Scirpus atrovirens

Black grass
Juncus gerardi

Cranberry
Vaccinium macrocarpon

Seaside gerardia
Gerardia maritima

Thread-leaved sundew
Drosera filiformis

Roadside—Borderline

9

Typical roadside flora
Black-pine scrub oak, cedar, shrubs, grasses
Route 27 east, Napeague, Amagansett, Long Island
New York

Roadside—Borderline

Introduction to Roadside Zone

1. Reed—*Phragmites communis*
2. Poison ivy—*Rhus radicans*
3. Bouncing bet or soapwort—*Saponaria officinalis*
4. Blue-eyed grass—*Sisyrinchium arenicola*
5. Sheep or common sorrel—*Rumex acetosella*
6. Yarrow—*Achillea millifolium*
7. Queen Anne's lace—*Daucus carota*
8. Butter-and-eggs—*Linaria vulgaris*
9. Charlock—*Brassica* species
10. Chicory (blue-sailors)—*Cichorium intybus*

Reed
Phragmites communis

Poison ivy
Rhus radicans

Bouncing bet or soapwort
Saponaria officinalis

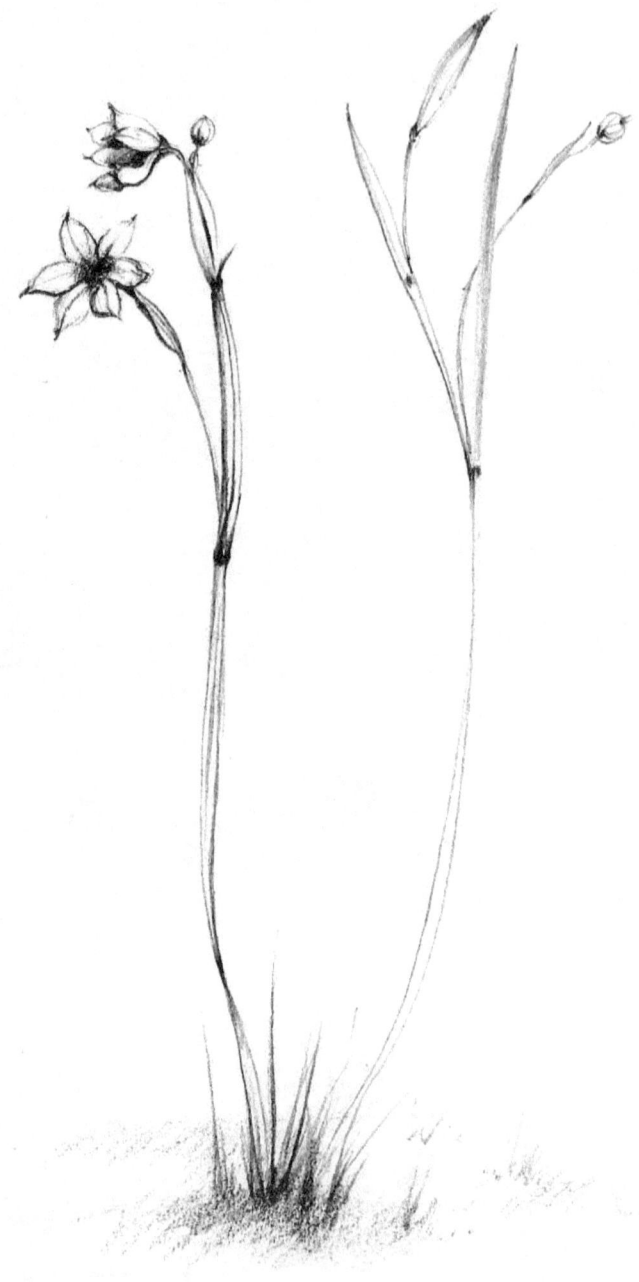

Blue-eyed grass
Sisyrinchium arenicola

Sheep sorrel
Rumex acetosella

Yarrow
Achillea millefolium

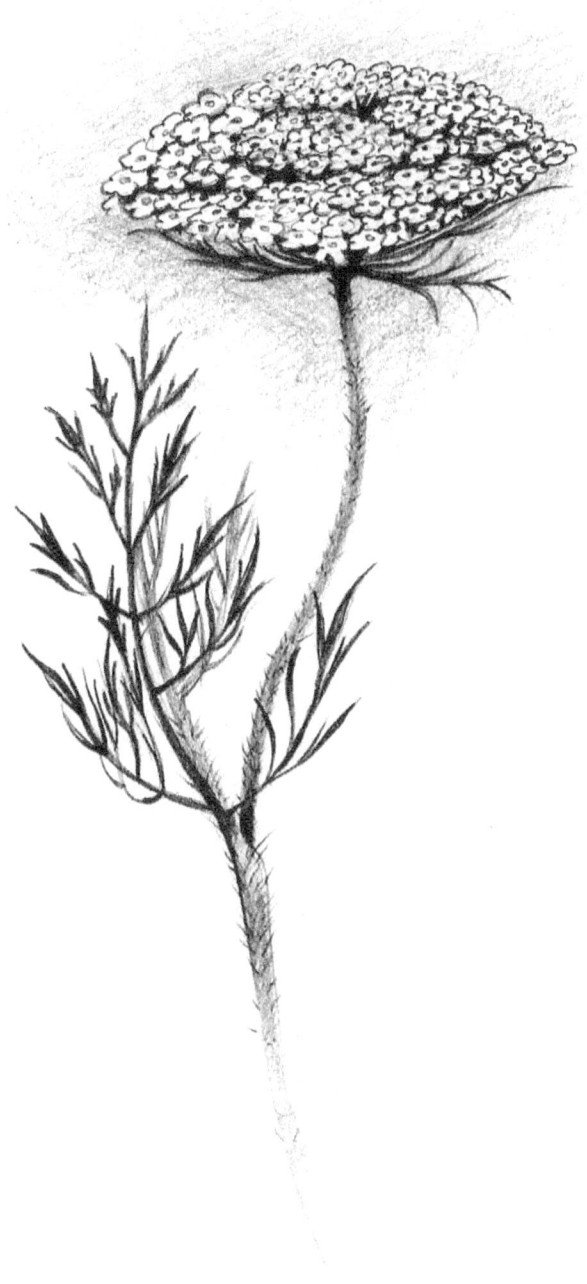

Queen Anne's lace
Daucus carota

Butter-and-eggs
Linaria vulgaris

Charlock
Brassica species—mustard family

Chicory (blue-sailors)
Cichorium intybus

References

Britton, Nathaniel Lord, and Addison Brown. *An Illustrated Flora of the Northern United States and Canada*. New York: Dover Publications Inc.

Gosner, Kenneth L. *A Field Guide to the Atlantic Seashore*. Boston, Massachusetts: Houghton Mifflin Co.

Hay, John. *Sandy Shore*. Chatham, Massachusetts: The Chatham Press Inc.

Kingsbury, John. *Seaweeds of Cape Cod and the Islands*. Chatham, Massachusetts: The Chatam Press Inc.

Merritt, L. Fernald. *Gray's Manual of Botany*. 8th ed. New York: American Book Co.

Minardi, Anthony S. *The Wetlands: Field Guide*. Indiana: Xlibris Publishing Co.

Miner, Roy Waldo. *Field Book of Seashore Life*. New York: G. P. Putnam's & Sons.

Norman, Marcia. *Beachcomber's Botany*. Chatam, Massachusetts: Chatam Conservation Foundation Inc.

Perry, Frances, ed. *Simon & Schuster's Complete Guide to Plants and Flowers*. New York.

McKenny, Margaret, and Roger Tory Peterson. *A Field Guide to Wild Flowers*. Boston, Massachusetts: Houghton Mifflin Co.

Index

B

barnacles, 24
barrier beach, xiii
black clam, 19
blue-sailors, 80
bog zone, 70
bouncing bet, 80, 83
Brassica species, 80, 89

C

charlock, 89
clamshell, 19, 25
coral seaweed, 3, 12

D

dulse, 3, 10

E

earthstar, 49
edible mussel, 24

F

floras, ix, xiii
   roadside, 79

H

halberd-leaved orache, 35

I

Iceland moss, 49
intertidal zone, 5
invertebrates, 15, 19
Irish moss, 3, 8

J

Jonah crab, 19, 23

K

kelp, 3–4
knotted wrack, 3

M

Moon snail egg casing, 25

N

northern starfish, 19, 26

Q

Queen Anne's lace, 80, 87

R

rockweed, 3, 7

S

salt spray rose, 49, 61
sand dollar, 22
sea clam, 4–5, 13, 19
sea clam substrate, 13
sea lettuce, 3, 13
sea scallop, 22
sickle-leaved golden aster, 57
soapwort, 83
sponge seaweed, 3, 6
surf clam, 20
surf zone, 1, 15, 17, 19
swale zone, 49

www.ingramcontent.com/pod-product-compliance
Lightning Source LLC
Chambersburg PA
CBHW030855180526
45163CB00004B/1581